# Preface.

The necessity and importance of the standardization of electrical apparatus was recognized in Germany as early as 1894 and the first rules („Sicherheitsvorschriften für elektrische Starkstromanlagen gegen Feuersgefahr") came into force in that country in 1895.

In the U. S. of America the first discussion on Standardization of Generators, Motors and Transformers took place in 1898 which resulted in the appointment of a Committee and the subsequent acceptance of the rules proposed by it.

In England the British Engineering Standards Association was formed in 1901. In connection with the British Standards an explanatory note appears necessary: The B.E.S.A.'s rules have for many years been the only generally accepted Standards. Since 1913 the British Electrical and Allied Manufacturers Association, representing the most important powerful British manufacturing firms, have issued Standardization Rules of their own which have attained considerable commercial importance. A special edition of these rules has been issued for export work which on the whole are guided by ideas similar to those embodied in the B.E.S.A. rules. As far as can be gleaned from the article in the "Electrical Review" (Vol. 86 Nr. 2, 216, April 16th 1920) it is intended to publish a new revision of the B.E.S.A.'s rules which will probably contain some of the recommendations of the B.E.A.M.A., so that this will probably mean the return to one single system of Standards for Britain.

The B.E.A.M.A.'s rules being provisional and less important than the B.E.S.A.'s rules are treated in separate Appendices.

In this connection the International Electrotechnical Commission Rules must also be mentioned which were adopted at the Plenary Meeting held in London 1919. It must be hoped that some day an international agreement on the most important points of Standardization of Electrical Machinery will be arrived at.

The I.E.C. Rules have not been included in the comparison because in practice they do not form yet the basis for important contract specifications for which as a rule one or the other national system of standards is still preferred. These rules are also very similar to the A.I.E.E. rules except in less important details.

In December 1922 a meeting of the I.E.C.'s Advisory Committee took place in Geneva. The main topic was the question of rating. After considerable discussion, a proposal on the lines desired by the

British delegates was agreed to for submission to the National Committees to the effect that overload ratings within the I.E.C. maximum continuous rating be recognised and that, for *general industrial machines (the class to be defined later)* where overloads have to be provided, the overload only to be applied under such conditions of air temperature as will not cause the limits of maximum temperature laid down by the I.E.C. to be exceeded, viz. $50^0$ C for Class A Insulation measured by thermometer at a maximum ambient air temperature of $40^0$ C.

This means clearly that the overload can only be applied at ambient air temperatures below $40^0$ C.

As far as Britain is concerned *Report No. 168—1923 of the B.E.S.A.* contains the *Standard specification for Industrial Motors and Generators* i. e. *Machines from 1 B.H-P. or kVA upwards with Class A Insulation wound for voltages not exceeding 7000 Volts having both Continuous- and Short-Time Ratings.* — This report came to hand only late but it was possible to include *the principal requirements which are marked thus:* 〉〈

In this connection it is important to note that the American delegates have declared that they object to the Geneva proposals. Under these circumstances an international agreement on this important question does not appear to be within easy reach.

In the various countries quite a number of revisions appeared necessary which had to take into consideration not only the enormous strides made by the different branches of Electrical Engineering, but also the experience gained in the application of the rules.

The study of the latest editions of the Standards of the three countries mentioned above, (the inclusion of other countries would not have been conducive to lucidity), shows that the views adopted on the principal points are related closely enough to allow a proper comparison to be made.

The purpose of this booklet is to enable the engineer or merchant buying from or selling to foreign countries, to inform himself *quickly* on any of the principal points telling in the design and performance of electrical machinery which he may be called upon to judge.

It is hoped, however, that it will also be of some use to the student and even to the consulting engineer in framing specifications.

For manufacturers and in cases of disputes, where legal decisions are involved, it will always be necessary to fall back on the original wording of the rules published by the bodies mentioned on the frontispiece; even in cases like this, the collation with the practice adopted in another country may afford guidance for the settlement of disputed points, on which the rules of the respective country might not prove precise enough.

Charlottenburg-Westend 1923.

**F. N.**

# COMPARISON OF PRINCIPAL POINTS OF STANDARDS FOR ELECTRICAL MACHINERY

(ROTATING MACHINES AND TRANSFORMERS)

BY

DIPL.-ING. FRIEDRICH NETTEL

CHARLOTTENBURG

Standards compared:

*GERMANY:*

Verband Deutscher Elektrotechniker (VDE)
a) Regeln für die Bewertung und Prüfung von elektrischen Maschinen. R. E. M. 1923
b) Regeln für die Bewertung und Prüfung von Transformatoren. R. E. T. 1923

*BRITAIN:*

1. British Engineering Standards Committee. (B. E. S. A.):
a) Standardization Rules for Electrical Machinery No. 72—1917.
b) British Standard Specification for the Electrical Performance of Industrial Motors and Generators with Class A Insulation. No. 168 — 1923.
2. British Electrical and Allied Manufactures Association. (B.E.A.M.A.):
Standardization Rules for Electrical Machinery. Fourth Edition 1920. Provisionally adopted.

*U. S. A.*

Standards of the American Institute of Electrical Engineers (AIEE) 1922 Revision

(Excluding Railway- and Tramway Motors and non-stationary Railway Transformers.)

Springer-Verlag Berlin Heidelberg GmbH 1923

ALL RIGHTS RESERVED

ISBN 978-3-662-24405-0      ISBN 978-3-662-26529-1 (eBook)
DOI 10.1007/978-3-662-26529-1

Copyright 1923 by Springer-Verlag Berlin Heidelberg
Originally published by Julius Springer in Berlin in 1923.

# Contents.

## Section I.
## Rotating Machines, excluding Railway- and Tramway-Motors.

|  | page |
|---|---|
| Standard pressures and Frequencies | 7 |
| Rating-Classification | 8 |
| Heating | 9 |
|    a) Standard- and max. permissible Temperatures of cooling media | 9 |
|    b) Determination of End Temperatures and Temperature Rises. Heating Test | 9 |
|    c) Method of measurement of Temperature of the cooling media | 10 |
|    d) Method of measurement of Temperature Rises | 10 |
|    e) Correction of Temperature Rises for Altitude | 11 |
|    f) Permissible Temperature Rises | 12 |
| High-Pressure Tests and Insulation Resistance | 14 |
|    1. High-Pressure Test | 14 |
|    2. Surge Test (Germany only) | 16 |
|    3. Winding Test and Insulation Resistance Test | 17 |
| Efficiency | 20 |
|    1. General | 20 |
|    2. Determination of Losses for Conventional Efficiency Test | 22 |
| Inherent Regulation | 25 |
| Overspeed Test | 26 |
| Tolerances | 27 |

### Appendix to Section I.

British Electrical and Allied Manufacturers' Association Rules (B.E.A.M.A.-Rules) . . . . . . . . . . . . . . . . . . . . . . . . . . . . . . . . . . 28

## Section II.
## Transformers.

| | |
|---|---|
| Rating-Classification | 31 |
| Heating | 33 |
|    a) Standard and max. permissible Temperatures of cooling media | 33 |
|    b) Determination of End-Temperatures and Temperature Rises | 33 |
|    c) Methods of measurement of Temperature of cooling media | 34 |
|    d) Methods of measurement of Temperature Rises | 35 |
|    e) Correction of Temperature Rises for Altitude | 35 |
|    f) Permissible Temperature Rises | 36 |

## Contents.

| | page |
|---|---|
| High-Pressure Test and Insulation Resistance | 38 |
|   1. High-Pressure Test | 38 |
|   2. Surge Test (Germany only) | 40 |
|   3. Winding Test and Insulating Resistance Test | 40 |
| Working in parallel (Germany only) | 40 |
| Overloads | 40 |
| Efficiency, Transformer Losses | 41 |
|   1. General | 41 |
|   2. Classification of Losses and their measurement | 41 |
| Inherent Regulation | 41 |
| Tolerances | 41 |

### Appendix to Section II.

British Electrical and Allied Manufacturers' Association Rules (B.E.A.M.A.-Rules) . . . . . . . . . . . . . . . . . . . . . . . . . . 42

# SECTION I.

# Rotating Machines.

**Standard pressures and Frequencies and Classification of Rating.**

|  | Germany | | Britain | | | U. S. A. |
|---|---|---|---|---|---|---|
|  | Generators | Motors | Motors | Consumer | Station | |
| **1. D. C. Generators and Motors. Volts.** | | | | | | |
|  | 115 | 110 | 110 | ⟩ For industrial machines ⟨ | | |
|  | 230 | 220 | 220 | 220 | 242 | |
|  | 460 | 440 | 440 | 440 | 484 | none |
|  |  |  | 460 |  |  | |
|  |  |  | 500 |  |  | |
| **2. Three-phase A. C. Generators and Motors. Volts.** | | | | | | |
|  | 130 | 125 |  | ⟩ For industrial machines ⟨ | | |
|  | 230 | 220 | 220 | 240 | 264 | |
|  | 400 | 380 | 400 | 416 | 457 | |
|  | 525 | 500 | 500 |  |  | |
|  | 3150 | 3000 |  | 3000 | 3300 | none |
|  | 5250 | 5000 |  | 6000 | 6600 | |
|  | 6300 | 6000 |  | (10000) | (11000) | |
|  | 10500 | 10000 |  |  |  | |
|  | 15750 | 15000 |  |  |  | |
| **3. Frequencies. Cycles per second.** | | | | | | |
|  | 50 | | 50 | | | 60 (usual) |

## Rating - Classification.

| Germany | Britain | U. S. A. |
|---|---|---|
| a) *Continuous Rating.* The machine shall give its rated output for any desired period without exceeding the limits specified for observable temperature and temperature rise.<br><br>b) *Short-time Rating.* The machine shall give its rated output for the specified period without etc. (as above). | a) *The British Standard for Continuous Rating* is the output at which a machine can work for an unlimited period and comply with these rules.<br><br>b) *Standards for Short-time Rating.*<br>1. One hour Rating. The output at which a machine can work for one hour and comply with the rules.<br>2. One half-hour Rating. The output at which a machine can work for one half-hour and comply with the rules. | a) *Continuous Rating.* A machine rated for cont. service shall be able to operate continuously at its rated output without exceeding any of the rules.<br><br>b) *Short-time Rating.* A machine rated for short-time service shall be able to operate at its rated load for the time specified in each case without exceeding etc. (as above). Standard Short-time ratings: 5, 10, 15, 30, 60, 120 minutes. |
| c) *Duty-cycle Rating.* Periods under pressure are followed by periods during which the machine is switched off the total duration of the cycle being max. 10 minutes.<br>The intermittency viz. the ratio of the period during which the machine is under current to the duration of the cycle characterizes the duty-cycle. The machine shall be able to work on a uniform duty-cycle service without the specified temp. limits being exceeded. Standard values for intermittency: 15, 20, 40 $\%$. | c) *Duty-cycle Rating* (in special cases) is the continuous rating which is the thermal equivalent of the cycle of duty specified. | c) *Duty-Cycle Rating.* For purposes of rating either a continuous or a short-time equivalent load may be selected, which shall simulate as nearly as possible the thermal conditions of the actual duty-cycle.<br><br>d) *Nominal Rating.* For Railway substation machines see as for transformers page 32. |

## Heating.

| Germany | Britain | U. S. A. |
|---|---|---|
| **a) Standard- and max. permissible Temperatures of cooling Media.** | | |
| Normal: 20°C <br> *) Highest permissible: 35°C | none specified <br> 40°C (air) <br> ⟩ 30°C (air)**⟨ | 25°C (water) <br> 40°C (air) |
| *) on this Temperature the End Temperatures on page 12 are based. <br> **) ⟩ when full amount of sustained overload specified on page 19 is required. ⟨ | | |
| **b) Determination of End Temperatures and Temperature Rises. Heating Test.** | | |
| The heating test is to be carried out at rated load | The temperature test shall be carried out at rated output | Whatever method of temperature measurement (see Section d) be employed it is required that: |
| 1) *for machines with continuous rating* until the temperature does not rise noticeably but not longer than 10 hours. | ⟩ 1) *for machines with continuous rating* until sufficient evidence is available to show that the max. temperature and temp. rise would not exceed the requirements of the rules if the test were prolonged until a steady final temperature were reached. | 1) The temperature test of a machine for *continuous service* shall be continued until sufficient evidence is available to show that the maximum temperature and temperature-rise would not exceed the requirements of the rules, should the test be prolonged until the attainment of a steady final temperature. |
| 2) for machines with *short-time rating* for the specified time, the testing being started with machine cold or when the Temperature of any part of the winding is not higher than 3°C than the Temperature of the cooling medium. <br><br> 3) for machines for *duty-cycle rating* until the temperature does not rise noticeably. <br> *The temperature is considered constant* if the rate of increase does not exceed 2°C per hour. | 2) *for machines with short-time rating* for the time required by the rating. ⟨ <br><br> **Note:** The temperature test shall be carried at any convenient air temperature not exceeding 40°C. | 2) The duration of test of a machine with a *short-time rating* shall be the time required by the rating. It shall commence only when the windings and other parts of the machine are within 5°C of the ambient Temperature. <br><br> **Note:** A machine may be tested at any convenient ambient temperature preferably not below 10°C. |

Section I.

Continuation: **Heating.**

| Germany | Britain | U. S. A. |
|---|---|---|
| \) c) **Methods of measurement of Temperature of the cooling media.** | | |
| *Same as Britain* but generally valid for "cooling media". <br><br> With large machines having parts below the floor-line it is admissible for test purposes to make the pit-temperature equal to the temperature outside of the pit. <br><br> For machines with *forced draught- and water-cooling* see Transformers Note page 34. | 〉 The air temperature shall be taken as the mean of the readings *of the thermometers* taken at equal intervals of time during the last quarter of the duration of the test. — For a pipe ventilated or forced draught machine, the air temperature is to be measured by a thermometer placed in the current of incoming air. 〈 | *Same as Britain* but for *Forced draught machines* a weighted mean shall be employed, a weight of four being given to the temp. of the circulating air and a weight of one to the room air. — Where machines are *partly below the floor line* the temp. of the rotor shall be referred to a weighted mean of the pit- and room temp. relative to the rotor portions in and above the pit. Stator parts in the pit shall be referred to the temp. in the pit. |
| d) **Methods of measurement of Temperature rises.** | | |
| The Temperature of windings is defined as *the higher of* the *two values* as below: <br><br> 1. *Mean Temperature rise by the Resistance Method.* <br><br> 2. *Temperature rise of the hottest accessible spot measured by Thermometer* (mercury or alcohol). Where Resistance Method is not applicable the Thermometer Method shall be used alone. | 〉 1. *Thermometer Method.* <br> Several thermometers (mercury or alcohol) shall be applied to the hottest accessible stationary parts during the test period and other thermometers to the rotating parts as soon as the machine is stopped †). 〈 <br><br> 2. *Resistance Method.* <br> The temperature rise shall be calculated from the resistances hot and cold, and the initial temperature (by thermometer) of the winding itself. Thermometer measurements shall also be made to ascertain if there is any higher observable temperature. If such is found it shall be adopted as the maximum observable temperature. | 1. *Thermometer Method.* <br> Mercury, alcohol, resistance or thermocouple thermometers applied to the hottest accessible part of the completed machine. <br><br> 2. *Resistance Method.* <br> Consists in the measurement of the temperature rise of windings by their increase of resistance. In the application of this method thermometer measurements shall also be made whenever practicable without dismantling the machine. The higher temperature of the two shall be taken as "observable" temperature. |

## Continuation: **Heating.**
### d) Methods of measurement of Temperature rises.

| Germany | Britain | U. S. A. |
|---|---|---|
| (Temp. Coefficient of copper = 235.) | 3. *Embedded Temperature Detector Method* same as U. S. A. but when specified with the enquiry only ††). Compliance with Methods 1 & 2 is however decisive. (Temp. Coeffic. of copper = 234,5.)<br><br>**Special cases:**<br>*D. C. Motors and Generators.*<br>Field coils Method 2 *or* Method 1 and 5°C lower Temp. rise.<br>*Synchronous Motors and Generators.*<br>Field windings: Method 2.<br>Stator: Method 2 *or* Method 1 and with 5°C lower temp. rise\*).<br>*Induction Motor or Generator.*<br>Stator & Rotor Method 2 *or* Method 1 with 5°C lower temp. rise\*).<br>\*) For Machines under 5000 Volts Method 1 alone is also permissible.<br>**Note:** †) ⟩ For industrial machines thermometer Method is used only. ⟨<br>††) Method 3 shall be employed for machines over 3000 kW for rated pressures exceeding 3300 Volts | 3. *Embedded Temperature Detector Method* consists in the measurement of the temperature rise by thermocouple or by resistance temperature detectors, located as nearly as possible at the estimated hottest spot. When required to be checked by method 2. (Temp. Coefficient of copper = 234,5.)<br><br>**Note:** Method 2 shall not be used for circuits of low resistance (other than transformer windings) as interpole windings. Method 3 should be applied to all Stators of machines having a width of 50 cm or over. It should also be applied to all machines of 5000 Volts or more, if rated over 500 kVA regardless of core width. |

### e) Correction of Temperature Rises for Altitude.

| | | |
|---|---|---|
| The rules hold good for service at altitudes *up to 1000 m* over sea level. If a machine will have to work in high altitudes this must be specially stated. | For a machine intended for service at an altitude over 1000 m (3300 ft.) the observable temperature rise if tested near sea level shall be reduced $2^1/_2$ *per cent for each 1000 feet.*<br>⟩ For industrial machines as above but reduction $1^1/_2$ *per cent for each 1000 feet.*<br>Machines for service at altitudes above 10000 feet are not considered standard. ⟨ | Generally the height above sea level at which a machine is intended for work is assumed not to exceed 1000 m. It is recommended that when a machine is intended to work at altitudes above 1000 m the permissible temperature rise at sea level shall be reduced by *1 per cent for each 100 m* by which the altitude exceeds 1000 m. |

## f) Permissible Temperature Rises (ET = End Temperature; TR = Temp. Rise).

### Germany

| Item | Class of Insulation | Part of machine | ET °C | TR °C |
|---|---|---|---|---|
| 1 | Class I: cotton silk paper — Not impregnated | A.C. Stator windings in slots | 75 | 40 |
| 2 | Class I: cotton silk paper — impregnated | All other windings except Item 9 & 10 | 85 | 50 |
| 3 | Class II: cotton silk paper impregnated (not oil-immers.) | as Item 1 | 85 | 50 |
| 4 | Class II (cont.) | as Item 2 | 95 | 60 |
| 5 | Class III: cotton silk paper in compound | All windings except Items 9 & 10 | 95 | 60 |
| 6 | Class IV: Enamelled wire | as Item 5 | 95 | 60 |
| 7 | Class V: Mica, Asbestos presparates if binding matter may be destroyed without impairing the insulat. | as Item 5 | 115 | 80 |
| 8 | Class VI: Raw Mica, Porcelain, fire proof materials | as Item 5 | Restricted by influence on adjacent parts only | |

### Britain — For pressures up to 5000 > 7000 (Volts*)

| Line | Class of Insulation | Part of machine | ET °C | TR °C | Method |
|---|---|---|---|---|---|
| 1 | Class O — same as class I Germany | A.C. windings in slots | 85 | 45 | see Remarks under d) |
| 2 | Class O — same as class I Germany | Rot. armatures with commutat. | 80 | 40 | |
| 3 | Class O — same as class I Germany | Stat. and rotat. D.C. field coils | 85 | 45 | |
| 4 | Class A — same as Germany class II also oil-immers. | as line 1 | 95 | 55 | |
| 5 | Class A | as line 2 | 90 | 50 | |
| 6 | Class A | as line 3 | 95 | 55 | |
| 6a | Class A — Industrial machines only | not tot. enclosed | >40 | | |
| 7 | | tot. enclosed | | >50 | |
| 8 | | short timerated | | >50 | |
| 9 | Enamelled wire | as line 1 | 95 | 55 | |
| 10 | Enamelled wire | as line 2 | 90 | 50 | |
| 11 | Enamelled wire | as line 3 | 95 | 55 | |
| 10 | Class B — Mica, Asbestos and similar material | as line 1 | 115 | 75 | |
| 11 | Class B | as line 2 | 110 | 70 | |
| 12 | Class B | as line 3 | 115 | 75 | |
| 13 | Class C, fire proof and refractory material | — | no limits yet specified | | |

### U.S.A. — Windings other than single-layer field windings and short circuited windings stated below

| Method of measurem. | Class of Insulation | ET °C | TR °C |
|---|---|---|---|
| 1 | Class O — same as Germany Class I | 75 | 35 |
| 2 | Class O — same as Germany Class I | 80 | 40 |
| 1 | Class A — same as Germany class II also when oil-immersed | 90 | 50 |
| 2 | Class A | 95 | 55 |
| 1 | Class A — Enamelled wire | 90 | 50 |
| 2 | Class B — same as Germany class V | 95 | 55 |
| 1 | Class B | 110 | 70 |
| 2 | Class B | 115 | 75 |
| — | Class C — same as Germany class VI | no limits yet specified | |

#### Open Types**

| ET °C | TR °C |
|---|---|

## Rotating Machines.

| Item | Part of machine | Classes I–VI (No temperature detectors specified) | | Insulation (Class O) Class A | | Class B | |
|---|---|---|---|---|---|---|---|
| | | | | ET | TR | ET | TR |
| 1 | single-layer field coils (on stator and rotor) with exposed surfaces*) | | | (85) 100 | (45) 60 | 120 | 80 |
| 2 | Other Rotor field windings | | | (80) 95 | (40) 55 | 115 | 75 |
| 1 | Rotor windings in slots | | | (75) 90 | (35) 50 | 110 | 70 |
| 2 | Short circ. windings insulated on stat. and rotat. | | | (80) 95 | (40) 55 | 115 | 75 |
| 1 | Cores and uninsul. short circ windings | | | (85) 100 | (45) 60 | 120 | 80 |
| 1 | Commutators, collector rings, bare metallic surfaces | | | ET 115 | TR — | ET 115 | TR 75 |
| | | | | as Britain line 15 | | | |
| — | Placing of temperature detector | Method | | | | | |
| 3 | Windings with 2 coil-sides per slot. Det. between top- and bottom coil-sides and between coil-sides and core | | | (85) 100 | (45) 60 | 120 | 80 |
| 3 | For windings with 1 coil-side per slot and detectors between coil-side and core and coil-side and wedge | | | (80) 95 | (40) 55 | 115 | 75 |
| 9 | Single-layer field coils with exposed surfaces | 100 | 65 | | | | |
| 10 | Continuously short circuited windings | 5°C higher than Items 1–7 | as Item 8 | | | | |
| 11 | uninsulated | — | as Items 1–7 | | | | |
| 12 | Iron core without embedded windings | — | as Item 8 | | | | |
| 13 | Iron core with embedded windings | — | as Items 1–7 | | | | |
| 14 | Commutator and sliprings | — | 95 | 60 | insulated 100 | 60 | |
| | | | | | uninsulated Restricted by influence on adjacent parts only | | |
| 15 | Bearings | — | 80 | 45 | ⟩ as line 15 ⟨ | | |
| 16 | All other parts | — | as Item 8 | | | | |
| 16 | Iron core without embedded windings | — | | Iron core without embedded windings | | | |
| 17 | Iron cores with embedded windings | — | | as lines 1–12 ⟩ 6a ⟨ | | | |
| 18 | Commutators and sliprings | — | | 50 ⟩ 45 ⟨ ... 90 | | | |
| — | Placing of temperature detector | Method | | Insulation Class A  ET TR | | Class B ET TR | |
| 19 | Windings with 2 coil-sides in one slot | 3 | | 5°C more than lines 4–6 | | lines 10–12 approx. same as lines | |
| 20 | Between coil-side and core below 5000 Volts | 3 | | Same as line 20 | | 4–6   10–12 | |
| 21 | Same as line 20 but over 5000 Volts | 3 | | 1°C more for each 1000 Volts over 5000 Volts than lines 4–6 | | 10–12 | |

No temperature detectors specified.

Note. Method of measurement:  Items 1—9. Method 1. Checked by 2
      "   10—16.   "   2.

*) When thermometers are applied direct to bare winding surface the Temp. Rise shall be 10°C higher than given by method 1.

**) For *enclosed machines* the Temp. Rise shall be taken 5°C higher than the values for all windings except short circuited insulated windings.

*) Motors and generators *over 5000 V* 1½°C less for each 1000 V or part thereof by which the rated pressure exceeds 5000 V.

Data marked thus ⟩ ⟨ refer (also) to industrial machines from 1 B.H.P. kW or kVA upwards, see Preface.

## 1. High-Pressure Tests.

| | Germany | Britain | U.S.A. |
|---|---|---|---|
| Nature of Tests | 1. High-Pressure Test<br>2. Surge-Test for windings over 2500 V.<br>3. Test on Turns* | 1. High-Pressure Test<br>2. Insulation Resistance*<br>Data marked thus ⟩⟨ refer (also) to industrial machines; see Preface. | 1. Test of Dielectric Strength<br>2. Insulation Resistance |
| | *1. High-Pressure Test* | *1. High-Pressure Test* | *1. Test of Dielectric Strength* |
| Wave form. | sinusoidal | sinusoidal | sinusoidal |
| Frequency | rated, or 50 cycles p. sec. | rated in general any between 25 and 100 per sec. | not less than rated frequency; 60 cycles per sec. recommended |
| Duration | one minute | one minute | one minute |
| Temperature during Test | working Temperature | working Temperature | working Temperature |

### Winding

**Germany**

| Item | Rated Output Rated terminal pressure E | Test pressure Volts | min. Volts |
|---|---|---|---|
| a | Output smaller than 500 Watts | 3 E | 2 E + 500 |
| b | Output larger than 500 W. E up to 5000 V. | 3 E | 2 E + 1000 |
| c | E over 5000 V. | 2 E + 5000 | — |

(all windings except items e—f)

**Britain**

| Item | Part of Machine | Rated Output | Test pressure Volts | min./max. Volts |
|---|---|---|---|---|
| a | all Machines | below 746 Watts or 1 H.P. | 2 E + 500 | — |
| b | all Machines | of 746 Watts or 1 H.P. and over | ⟩ 2 E + 1000 ⟨ | ⟩ 2000 above 3 H.P. ⟨ (only) |
| c | all Machines in service (i. e. not new) | below 746 Watts | 1,5 E + 375 | — |
| | | of and over 746 Watts | 1,5 E + 750 | — |

**U.S.A.**

| Item | Winding | Rated Output Normal circuit voltage E | Test pressure Volts | min./max. Volts |
|---|---|---|---|---|
| | Small machines less than 25 V. | | 500 | — |
| a | Small Generators and Motors under | 660 Watts or 0,5 H.P. voltage under 275 V. to 25 V. | 900 | — |
| b | all Machines except as otherwise specified (see below) | | 2 E + 1000 | — |
| c | A.C. machines connected to permanently grounded single phase systems. (Not 3-phase with neutral grounded) | more than 300 Volts | 2,73 E' + 1000 E' voltage of circuit to ground | — |

# Rotating Machines.

| | Exciter-windings of Rotary Converter and Synchronous Motors | Separately excited rotating field windings (Exciter volts not over 750 <) | | Field windings of Synchronous Machines or Rotary Converters | | Field windings of A.C. Generators | | Field windings of Synchronous Machines including Motors and Converters | | Wound Rotors of Induction Motors | | Phase-wound Rotors of Induction Motors | |
|---|---|---|---|---|---|---|---|---|---|---|---|---|---|
| | | Exciter | | | | | | | | | | ordinary | 2 E + 1000 |
| d | with Exciter circuit always closed with or without starting from A.C. side | 3 E + 1000 | for starting with A.C. with fields closed-circuited | 10 E + 1000 | for starting with A.C. with fields shorts circuited | 10 E | for starting with A.C. with field shorts circuited | 10 E | min. 1500 max. 3500 | non-reversing according to output as above | 2 E + 500 > or 2 E + 1000 < | | |
| e | with Exciter winding sectionalized without or with starting from A.C. side | 10 E + 1000 | for starting with A.C. with break-up switch | same as field windings | for starting with A.C. with break-up switch | min. 1500 > 2000 < max. 3500 | for starting with A.C. with fields open circuited and sectionalized while starting | 5000 | | | reversing | > 4 E + 1000 < | reversing | 4 E + 1000 |
| f | with disconnectable exciter circuit without starting from A.C. side | 10 E + 1000 | for starting with A.C. without break-up switch and fields open | 10 E + 1000 | for starting with A.C. with break-up switch | min. 1500 > 2000 < max. 3500 | | 5000 | | | | | Standard machines produced in large quantities 2500 V. or less | 20 % higher than above |
| g | as Item f but with starting from A.C. side | 20 E + 1000 | | 5000 for Exciter Volts less than 250 > 275 < and above | for starting with A.C. fields open | 5000 for Exciter Volts less than 275 > and above | for starting with A.C. with fields open circuited and connected all in series while starting | 5000 for Exciter Volts less than 275 to 750 | 8000 for Exciter Volts 275 to 750 | | | | Assembled apparatus tested as electrical unit | 15 % lower than lowest individual test pressure |

**Note.** E represents:
1. The rated terminal pressure of the machine, with field windings the rated exciter pressure.
2. For different direct connected windings of one or more machines the highest possible pressure against the frame when one pole is grounded.
3. For phase-wound Rotor windings of asynchronous motors; non-reversing: the rotor pressure, reversing 1,5 times rotor pressure.
4. For machines with one rotor pole, permanently grounded 1,1 times rated pressure. Short circuited windings need not be tested.

\*) Sequence of Tests: 1, 2, 3.

**Note.** E represents:
a) The test pressure shall be based on the rated pressure or the highest pressure reached between any part of the winding and the frame, which ever is greater.
b) For a machine driven by water-wheel and exposed to possible excess pressure, pressure limiting devices should be employed otherwise test as under a) is necessary.

\*) Sequence: advisable as before 1.

**Note.** E represents:
The normal voltage of the circuit to which the machine is connected.
For Rotors of Induction Motors the normal induced voltage.

## 2. Surge Test (Germany only).

The surge test which serves to ensure the efficiency of the insulation against surges occurring under ordinary working conditions shall be made with the completely assembled machine on the factory test bed as far as possible in accordance with the diagrams of connections as shown below for Synchronous- and Asynchronous machines. (G = Generator, M = Motor.)

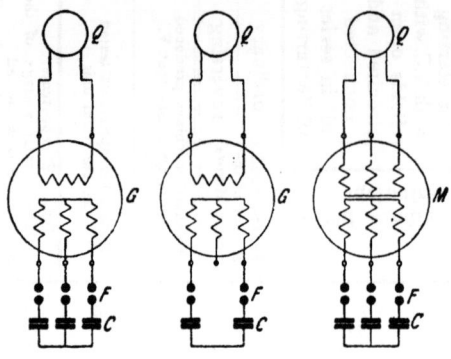

The machine on test shall be connected to cables or static condensers (C) over sphere spark-gaps (F) with solid spheres of at least 50 mm diameter.

The Test Capacitance of the cables or condensers shall be as follows:

| Rated voltage E kV | Capacitance in each phase min. $\mu$F. (Mikrofarads) |
|---|---|
| 2,5 to 6 | 0,05 |
| up to 15 | 0,02 |
| over 15 | 0,01 |

Each spark-gap shall be set for 1,1 times rated Voltage. The machine shall be excited by a d. c. supply to abt. 1,3 times rated voltage at rated speed or with 3-phase a. c. at rated frequency respectively. The spark-gaps shall be caused to break down in any manner (say by temporarily reducing or bridging of the gap) and the arcing maintained for a period of 10 secs. During the test the spark-gap shall be placed in an air current of a velocity of 3 m per sec.

Care shall be taken to make all connections for the purposes of this test as short as possible.

Polyphase machines may be tested in single phase connection. The connections shall in this case be changed as often as necessary so that every phase will be subjected to the surge test.

## 3. Test on Turns.   2. Insulation Resistance Test.

| Germany | | Britain | U. S. A. |
|---|---|---|---|
| The test on turns shall be made under no-load condition by increasing the supplied and generated voltage respectively (Motors & Generators) up to the values given below. The frequency and speed respectively may be increased correspondingly.<br><br>Duration of Test: 3 Minutes | | | The insulation resistance may afford a useful indication as to whether the machine is in suitable condition for application of the dielectric test. The insulation resistance test shall be made with all circuits of equal voltage above ground connected together. Circuits or groups of circuits of different voltage above ground shall be tested separately.<br><br>*Test pressure if possible D. C. 500 volts. The minimum value* of insulation resistance at operating temperature shall be: |
| Winding | Test voltage / Rated voltage | same as U. S. A. | |
| a  All windings except Item b | 1,3 | $\text{Megohms} = \dfrac{\text{Rated volts}}{\text{Rated kVA or HP} + 1000}$ | $\text{Megohms} = \dfrac{\text{voltage at terminals}}{\text{rating in kVA} + 1000}$ |
| b  Polyphase Windings with fixed (not disconnectable) connections between the different parts | 1,5 | | The formula applies only to dry apparatus not for oil-immersed types. |

| Rated voltage | Megohms | | |
|---|---|---|---|
| | 100 kVA | 1000 kVA | 10 000 kVA |
| 100 | 0,091 | 0,05 | — |
| 1000 | 0,91 | 0,50 | 0,091 |
| 10 000 | 9,1 | 5,0 | 0,91 |
| 100 000 | — | 50 | 9,1 |

## Overloads, Starting, Commutation Test.

| Germany | Britain | U. S. A. |
|---|---|---|
| All machines with *continuous rating* — must be capable of withstanding *1,5 times rated current for 2 Minutes*, following cont. full load run, without deleterious effect or resulting lasting deformation.<br><br>This test shall be carried out for<br><br>*Motors & Rotary Converters* at rated pressure; *for Generators* the pressure should be regulated as closely as possible to the rated pressure. (See Note below.)<br><br>*Motors* shall be required to develop the following *max. running torques* (or more) without stalling. | The machines shall be capable of withstanding on test the following *Excess currents* or torques in excess of those corresponding to their Brit. Stand. Rating the pressure being maintained as near the rated value as possible:<br><br>a) *Machines for continuous service.*<br>Excess Amperes:    *150 per cent*<br>Duration:    *15 seconds.*<br><br>b) *Induction Motors for continuous service.*<br>Excess torque without stalling:    *175 per cent* (except for abnormally low speed motors or high frequencies). | *Continuously rated machines* shall be required to carry *momentary loads of 150 per cent of the amperes* corresponding to the continuous rating keeping the rheostat for rated load excitation.<br><br>*Motors for continuous service* shall be required to develop *running torque at least 175%* of that corresponding to the running torque at their rated load without stalling.<br><br>*Machines with nominal rating.* See as for Transformers. page 32. |

| Rating: | max. Torque (% of rated) |
|---|---|
| Continuous Short-Time | 16 % |
| Intermittent | 20 % |

| Germany (cont.) | Britain (cont.) | U. S. A. (cont.) |
|---|---|---|
| A. C. Motors shall be required to develop a starting torque of at least 30% of the rated torque, when working with a suitable starter during the whole starting period in each position.<br><br>**Note:** Generators should be amply designed so as to enable them to generate rated pressure at rated speed, power factor and exciter voltage and at 125% of rated current. | c) *D. C. shunt wound Motors or squirrel cage induction motors for intermittent service.*<br>Excess torque when running:    *200 per cent*<br>Duration:    *30 seconds.*<br><br>d) *All other motors* for intermittent service shall have *starting torques equal to 200 per cent of rated torque.* | Obviously *dutycycle* machines must carry their peak load without stalling. |

## Overloads, Starting, Commutation Test.

| Rating | Motors BHP per 1000 Revs. p. M. not totally enclosed [totally enclosed] | Generators kW or kVA | Sustained Overload (Torque or current respectively) % | Sustained Overload Duration | Momentary Overload % | Momentary Overload Duration |
|---|---|---|---|---|---|---|
| ⟩ Britain: Industrial machines only. ⟨ †) | | | | | | |
| continuous | below 4 to 1 | below 3 to 1 | 25 [—] | ½ hour [—] | 50 [50] | 1 Min [15 secs] |
| continuous | 4 & upwards | 3 & upwards | | | 50 [50] | 1 Min [15 secs] |
| continuous | D.C. up to 150 H.P. incl. | — | 25 [—] | 2 hours [—] | 100 [75] | 15 secs [15 secs] |
| continuous | A.C. all sizes *) | — | | | 100 [75] | 15 secs [15 secs] |
| short t. | all sizes | all sizes | — [—] | — [—] | 100 [100] | 30 secs [30 secs] |

†) *not* for motors of abnormally low speeds or high frequency.
*) for single phase motors overloads are not yet specified.

| Germany | Britain | U. S. A. |
|---|---|---|
| **Commutation Test.** *Machines with Commutators* shall work practically sparkless from no-load to full load. *At overload test* they must commutate in such a way so that neither commutator nor brushes are injured. Machines with Commutators shall work practically sparkless with fixed brush setting as follows: | **Commutation Test.** ⟩ D. C. machines shall work throughout the range from no load to the highest momentary load specified with fixed brush setting. The operation must be practically sparkless from no load to full load and without injurious sparking up to the highest momentary load specified. ⟨ | **Commutation Limitation.** *Continuously rated machines* shall be required to commutate successfully *momentary loads* of 150 % *of the Amperes* corresponding to the continuous rating keeping the rheostat set for rated load excitation. *Machines for duty-cycle operation* shall commutate successfully under their specified conditions. Successful operation is such that neither brushes nor commutator are injured by the test. |

| without | with |
|---|---|
| auxiliary poles | |
| from 0,25 to 1,0 load | from no load to full load |

## Efficiency[*].

### 1) General

| Germany | U. S. A. |
|---|---|
| | |

*Recognized Efficiencies:*

| | |
|---|---|
| 1. Directly measured Efficiency $$= \frac{\text{Output}}{\text{Input}}.$$ 2. Conventional Efficiency (determined from losses). | 1. Directly measured Efficiency $$= \frac{\text{Output}}{\text{Input}}.$$ 2. Conventional Efficiency (determined from losses). |

*General conditions:*

| | |
|---|---|
| Unless otherwise specified the conventional efficiency is to be employed.  Efficiency figures shall correspond to normal working conditions unless stated otherwise. When using method 2 the resistances measured by D. C. shall be corrected to a *reference temperature of* $75^0 C$. For measurement of other losses no temperature correction shall be made. | Unless otherwise specified the conventioal efficiency is to be employed. Efficiency figures shall correspond to rated load, sine wave form, *temperature of reference* (at all loads) $75^0 C$ (tests shall preferably not be made at an ambient temperature of less than $15^0$ C). The efficiency of alternators shall be stated at the rated power factor. Sine wave shall be standard unless different wave form is inherent in the operation of the system. |

[*] Britain: Neither Definition nor Specification regarding measurement stated. 〉For industrial machines the declared (actual) efficiency shall include stray load losses estimated or ascertained as closely as possible.〈

## Efficiency.

### 1) General

| Germany | U. S. A. |
|---|---|
| *Miscellaneous losses:* | |
| All *losses due to auxiliary apparatus* — but these only — of the machine itself shall be charged against the machine efficiency. Such are:<br>a) *Losses in Regulating-, Series-, Calibrating-, Shunt-* and similar resistances, *Choking coils, Auxiliary transformers* etc. which are necessary for the regular working. (Exception see under c.)<br>b) *Exciter losses* in case of direct or indirect coupled exciter; not in case of separate excitation.<br>c) *Losses in booster* machines *of Rotary Converters*, if they form part of the converter proper but not the losses of the transformers and choking coils belonging to the converter. These losses shall be stated separately.<br>d) *Losses in bearings* supplied with the machine but not in bearings supplied by another firm.<br>e) The power consumption of the *ventilating blower* if this is part of the machine proper. The power consumption in case of *separate ventilating blower* as also that of *water and oilpumps* is not to be charged against the machine efficiency but shall be stated separately. | a) *Field-Rheostat Losses* shall be included in the generator losses where there is a field rheostat in series with the field magnets, even when the machine is separately excited.<br>b) *Ventilating Blower.* When a blower is supplied as part of a machine set, the power required to drive it shall be charged against the complete unit, but not against the machine alone.<br>c) *Other Auxiliary Apparatus* such as a separate exciter for a generator or motor, shall have its losses charged against the plant of which the generator and exciter are a part and not against the generator. An *exception* should be noted in the case of *Turbo-generator sets* with direct-connected exciters, in which case the losses in the exciter shall be charged against the generator. The actual energy of excitation and the field-rheostat losses, if any, shall be charged against the generator.<br>d) *For the Booster type of Synchronous Converters* where the booster forms an integral part of the unit its losses shall be included in the total converter losses. |

## 2. Determination of Losses for Conventional Efficiency Test.

References to Explanations given below. (Britain none specified.)

| Type of Machines | Standards | $I^2R$ Losses no-load (Shunt windings and separately excited circuits) | $I^2R$ Losses load (Armature and Series windings) | Bearing friction and windage | Brush friction of commutators and collector rings | Core loss | Brush contact $I^2R$ Loss | Stray Load Losses |
|---|---|---|---|---|---|---|---|---|
| D. C. Commutating Machines | G. | I a) | I b) | | 2 b, d) | | 3 | 4 a, b) |
| | U.S.A. | I a) | | II a, c) | III a) | IV a) | V | — |
| A. C. Commutating Machines | G. | I a) | I b) | | 2 a, d) | | 3 | — |
| | U.S.A. | I a) | | II a) | III a) | IV a) | V | — |
| Synchronous Motors and Generators | G. | I a) | I b, c) | | 2 c, d) | | 3 | I e $\alpha, \beta$) |
| | U.S.A. | I a) | | II a, c) | negligible except when armat. revolving | IV b) | negligible except when armat. revolving | VI a) |
| Synchronous Converters | G. | I a) | I d) | | 2 a, d) | | 3 | 4 c) |
| | U.S.A. | I a, c) | | II a) | III a) | IV a, b) | V | — |
| Induction Machines | G. | — | I c) | | 2 a) | | 3 if any | 4 d) |
| | U.S.A. | I a, b) | | II b) | III a) | IV a, c) | V if any | VI b) |
| Cascade Converters | G. only. | I a) | I d) | | 2 a, d) | | — | 4 c) |

### References:

| Germany (G.) | U. S. A. |
|---|---|

#### 1. $I^2R$ Losses.

**Germany:**

a) *$I^2R$ losses at no-load* shall be calculated from resistances measured by D. C.

b) *$I^2R$ losses due to load current* shall be calculated from resistances measured by D. C.

c) *For Induction Motors* the $I^2R$ losses in the secondary circuit should be calculated from the slip.

#### 1. $I^2R$ Losses.

**U.S.A.:**

a) *The $I^2R$ losses* shall be based upon the current and the measured resistance.

b) *The $I^2R$ loss* in the *Rotors of Polyphase Induction Motors* should be determined from the slip whenever the latter is accurately determinable using the following aquation:

$$\text{Rotor } I^2R \text{ loss} = \frac{\text{output} \times \text{slip}}{1 - \text{slip}}.$$

## Continuation: **Determination of Losses for Conventional Efficiency Test.**

| Germany | U. S. A. |
|---|---|
| d) *For Rotary (Synchronous) Converter* the $I^2R$ losses in the armature windings shall be *calculated* from those corresponding to its use as D. C. Generator by using the following factors: <br><br> <table><tr><td>Number of phases:</td><td>1</td><td>2</td><td>3</td><td>6</td><td>12</td></tr><tr><td>Number of sliprings:</td><td>2</td><td>4</td><td>3</td><td>6</td><td>12</td></tr><tr><td>Factors:</td><td>1,45</td><td>0,39</td><td>0,58</td><td>0,27</td><td>0,2</td></tr></table> <br> e) *Synchronous Motors and Generators.* Determination of *$I^2R$ plus stray load losses.* <br> α) Short-circuit method: The machine with short circuited armature winding shall be driven by a calibrated auxiliary motor at rated speed and excited so that the current is equal to rated current. The input *exclusive* of $I^2R$ loss at no load, bearing friction and windage, brush friction and core loss is considered equivalent *to $I^2R$ losses due to load plus stray load losses.* <br> β) Over-excitation method: The machine at no-load at rated pressure and frequency shall be overexcited so as to carry rated current. The input equivalent as under α). <br><br> **2. Bearing friction, Brush friction windage and core losses.** <br> **Motor method:** Machine shall be worked as motor at no load. <br><br> a) *A. C. machines* at rated pressure, rated frequency and at no load speed. <br> b) *D. C. machines* at rated pressure + or − the ohmic pressure drop and at rated speed. <br> c) *Synchronous machines* with excitation regulated so as to take the minimum current. <br> For a) — c): <br> The input minus $I^2R$ loss and excitation losses is considered equivalent to *bearing-, brush friction and windage losses as above plus core losses.* | In *large Slip-ring motors* in which the slip cannot be directly measured by loading, the rotor $I^2R$ loss shall be determined by direct resistance measurement; the rotor full-load current to be calculated by the following equation: <br> Current per ring $$= \frac{\text{watts output}}{\text{rotor volts at standstill} \times \sqrt{3} \times K}.$$ This applies to 3-phase motors; for 2-phase rotors use 2 instead of $\sqrt{3}$. For motors of 150 kW or larger $K = 0{,}95$. For smaller motors K decreases; no specific value stated. <br><br> c) *Synchronous Converters.* $I^2R$ losses in armature windings from losses corresponding to its use as D. C. generator by using suitable factors. <br><br><br> **II. Bearing friction and windage:** <br> a) *General:* Drive the machine from an independent motor, the output of which shall be suitably determined. The machine under test shall have its brushes removed and shall not be excited. The output represents the bearing friction and windage. <br> b) *Induction motors:* Losses as above may be measured by running motors free at the lowest voltage at which the motors will rotate continuously at approximately rated speed; the watts input minus $I^2R$ loss, under these conditions being taken as the friction and windage. <br> c) *For Engine-type Generators* losses as above are very small and shall be neglected. |

Continuation: **Determination of Losses for Conventional Efficiency Test.**

| Germany | U.S.A. |
|---|---|
| d) *Generator method:* The machine shall be driven at no-load and rated speed by a calibrated auxiliary motor and excited to rated pressure. The mechanical input minus excitation losses are equivalent to losses as under c). <br><br> Note: *For D. C. machines* plus or minus the ohmic pressure drop. | **III. Brush friction of Commutator and Collector Rings.** <br> *General:* Machine shall be driven as under II a) but with brushes in contact. The brush friction is equivalent to the difference of motor output obtained in test II a) and this test. <br><br> **IV. Core loss.** <br> a) *General:* Machine shall be driven as under III but excited so as to produce at the terminals a voltage corresponding the calculated internal voltage for the load under consideration. The core loss is equivalent to the difference of motor output obtained by this test and that in test III. <br><br> b) *Synchronous Machines:* The internal voltage shall be determined by correcting the terminal voltage for the resistance drop only. <br><br> c) *For Induction Motors* the core loss is determined by measuring the watts input when running free at rated voltage and frequency minus no-load copper loss, bearing friction and windage. |
| 3. **Brush Contact $I^2 R$ loss** is calculated by assuming the pressure drop in each brush as follows: <br> Carbon and graphite brushes   1 Volt. <br> Metal-graphite brushes   0,3 Volt. | **V. Brush Contact $I^2 R$ loss.** <br> Carbon and graphite brushes 1 Volt Standard (pigtails attached). <br> Carbon and graphite brushes 1,5 Volt Standard (pigtails not attached). <br> Metal-graphite brushes considered special. |
| 4. **Stray load losses.** See also 1 e $\alpha$) and $\beta$). Stray load losses for other types are determined in accordance with the following approximate values. The per cent values correspond to output in the case of generators, to input in the case of motors, to D. C. output in the case of rotary converters. It is assumed that they vary with the square of the current. <br> a) Compensated D.C. machines   0,5 % <br> b) Not compensated D. C. machines with or without commutating poles   1 % <br> c) Rotary converters   0,5 % <br> d) Asynchronous machines   0,5 % <br> e) Cascade Converters   1 % | **VI. Stray load losses** <br> a) *for Synchronous Machines* s. l. 1. are determined as under 1 e $\alpha$) (Germany) minus $I^2 R$ losses for polyphase generators and motors. For single-phase machines stray load losses are large but not yet specified. <br><br> b) *Induction Machines* with rotor removed, measure the input to the stator with different currents. Deduct $I^2 R$ loss determined from resistances. The difference represents *approximately* the stray load losses at the different currents. |

## Inherent Pressure Regulation

(to be measured at the temp. attained under normal operation).

| Line | Type of machine | Standard | Alteration of load condition for purposes of test*) | Regulation in % of rated voltage is determined as | General conditions during test ||| 
|---|---|---|---|---|---|---|---|
| | | | | | Speed or frequency | Excitation | Remarks |
| 1 | D. C. Generators a) shunt wound and | G. | from rated load to no-load | voltage rise | rated | for a) resistance in shunt field circuit constant | brushes in normal position |
| 2 | b) separately excited | U.S.A. | | | | for b) exciting current constant | — |
| 3 | D. C. Generators c) compound wound | G. | from rated load to no-load and from no-load to rated load | difference between highest and lowest voltage | rated | as line 1 | |
| 4 | (also series) | U.S.A. | | mean value of the two results | | as line 1 (constant resistance shunting series field) | |
| 5 | | G. | | voltage variation (shall not exceed 50% at P.F. = 0,8) | | | at specified P.F. |
| 6 | Synchronous Generators with direct connected exciter and separate excitation | Brit. | from rated load to no-load | voltage rise | rated | excitation current constant | Power Factor should be specified; if not P.F. assumed = 1 |
| 7 | | U.S.A. | | | | | |
| 8 | Synchronous and Cascade converters | G. | from rated load to no-load | voltage rise on D. C. side | rated | as line 1 | |
| 9 | Synchronous converters | U.S.A. | | | | | |
| | | G. | *) If Inherent Regulation cannot be determined by Load Method it can be calculated from the magnetic-test Curve. Reference temperature of resistances 75° C. |||||
| | | U.S.A. | **) As for Germany. For Synchronous Generators I.R. can also be determined from estimated zero Power Factor Curve. |||||
| | | Brit. | 〉For industrial machines: D.C. Generators as line 1. Synchr. Generators as line 6.〈 |||||

## Overspeed Test.
(Britain none specified.)

| Line | Type of machine | Germany | U.S.A. |
|---|---|---|---|
| | Duration: | 2 minutes | none specified |
| | | Test speed / Rated speed | Test speed / Rated speed |
| 1 | Generators except lines 2 & 3 | 1,2 | 1,25 |
| 2 | Generators driven by waterwheels | 1,8 | tested for max. runaway speed |
| 3 | Generators driven by steam turbines | 1,25 when equipped with emergency governor acting at 10% overspeed | 1,2 when equipped with emergency governor |
| 4 | Synchronous and Cascade Converters | 1,2 | — |
| 5 | Motors for constant speed | Test speed / no-load speed = 1,2 | 1,25 |
| 6 | Motors with several rated speed | Test speed / max. no-load speed = 1,2 | Test speed / max. speed (rated) = 1,25 |
| 7 | Motors with speed variation | 1,2 | 1,25 |
| 8 | Motors with series characteristic | Test speed / max. rated speed = 1,2; Test speed / rated speed = 1,5 | not specified |

## Rotating Machines.

## Tolerances.
(U.S.A. none specified.)

| Guarantees for: | Tolerances | | |
|---|---|---|---|
| | Germany | | Britain |
| Speed of Shunt wound Motors | Output<br>0 to 1,1 kW<br>1,1 to 11 kW<br>over 11 kW | ±% of rated<br>10<br>7,5<br>5 | ± 5% of speed at the temp. of full load |
| Speed of Series wound Motors | 0 to 1,1 kW<br>1,1 to 11 kW<br>over 11 kW | 15<br>10<br>7 | ± 7,5% of speed at the temp. of full load |
| Speed variation of D.C. Motors | 10% of specified variation | | — |
| Speed variation of Asynchronous machine | 20% of specified slip | | — |
| Inherent Pressure Regulation of Generators | 5% of rated pressure | | — |
| Inherent Pressure Regulation<br>of Rotary Converters<br>of Cascade Converters | ± 1% of rated pressure<br>± 3% „ „ „ | | —<br>— |
| Momentary short-circuit current of Synchronous Machines | 20% of specified value | | — |
| Stationary short-circuit current of Synchronous Machines | 15% of specified value | | — |
| Max. running torque of Motors | 10% of specified value | | — |
| Starting torque of Motors | 10% | | — |
| Efficiency $= \eta$ | $\frac{1-\eta}{10}$ rounded up to $\frac{1}{1000}$; but minimum 0,01 | | — |
| Power Factor $= \cos \varphi$ | $\frac{1-\cos\varphi}{6}$ rounded up to $\frac{1}{1000}$; but min. 0,02 | | — |
| Temperature Rise | — | | expressly none. |

## Appendix to Section I.

British Electrical and Allied Manufactures Association Rules.
Fourth Edition 1920.

| 1 | 2 | 3 | | | |
|---|---|---|---|---|---|
| **Standard pressures and Frequencies.** | **Temperature Test.** | **f) *Permissible Temperature Rises.*** | | | |
| | The temp. test shall be carried out at rated output for: | Insulation Part of machine | Ventilation | | |
| | | | unobstructed | partially obstructed | Totally enclosed |
| | | | T R °C | T R °C | T R °C |
| 1. *Generators:*<br>D. C. \| A. C.<br>230 \| 440<br>460 \| 550<br>525 \| 2200<br>\| 3000<br>\| 6600<br>\| 11000 | 1. *Machines with continuous rating* until the temperature rise is practically constant, that is until the rate of increase of temperature does not exceed 1°C per hour.<br><br>2. *Machines with short-time rating* for the period defined by the rating. | All windings whose insulation consists wholly or in important part, of cotton, paper, varnished cloth or similar materials. | 40 (therm.) | 47 (therm.) | 55 (therm.) |
| 2. *Motors:*<br>D. C. \| A. C.<br>110 \| 100<br>220 \| 200<br>440 \| 400<br>500 \| 500<br>at consumers terminals.<br><br>3. *Standard Highs pressure Systems for A. C.:*<br>2000, 3000, 6000, 10000, 20000 V.<br><br>4. *Frequencies:*<br>50 and 25.<br><br>**Classification of Rating:**<br>a) continuous<br>b) short-time | c) *Methods of measurement of Temperature of the cooling media* same as B.E.S.A. rules.<br><br>d) *Methods of measurement of Temperature rises.*<br>1. Thermometer Method (generally)<br>2. Resistance Method (for special cases). | Mica Asbestos Preparates | Higher values permissible (not specified) | | |
| **Heating.**<br>a) *Standard and max. permissible Temperatures of Cooling Media:*<br>Normal 25°C (air)<br>Max. 35°C (air).<br><br>b) *Determination of End-Temperatures* and *Temperature Rises* | e) *Correction of Temperature Rises for Altitude.*<br>Temp. Rise to be reduced $2\frac{1}{2}$ per cent for each 1000 feet. | Raw, Mica Porcelain, fire proof Materials | Higher values permissible (not specified) | | |

## Continuation: Appendix to Section I.

| | 4 | | | 5 | | | 6 |
|---|---|---|---|---|---|---|---|
| Insulation Part of machine | Ventilation | | | High Pressure Tests and Insulation Resistance. | | | Note: E represents: |
| | unobstructed | partially obstructed | Totally enclosed | Nature of tests etc. as B.E.S.A. rules!<br>A. High Pressure Test. | | | 1. The test must be based on the highest pressure to which the windings may be subjected. H. P. tests on field windings are to be based on the excitation pressure. |
| | T R °C | T R °C | T R °C | Rated terminal pressure E | Test pressure Volts | Min. Volts | |
| Core in which windings are embedded | 40 | 47 | 55 | Not more than 333 Volts | 1000 | — | 2. In the case of machines driven by waterwheels and exposed to runaway conditions or otherwise exposed to possible excess pressure it is recommended that pressure-limiting devices shall be provided, otherwise test as under 1) necessary. |
| Commutator and Sliprings | | 55 | | Above 333 but not more than 1500 Volts | 3 E | 1500 | |
| Alternator field coils | 55 (Res.) | | · | Above 1500 but not more than 2250 Volts | 4500 | — | |
| | | | | Above 2250 Volts | 2 E | — | B. Insulation Resistance Test. |
| Shunt field coils of D.C. machines | 60 (Res.) | 65 (Res.) | 70 (Res.) | Note: Field windings of synchronous machines intended to be started from the A. C. side are to be tested with unless the field windings are provided with a "breakup" switch, or will always be short circuited at starting | 5000 | — | In general the following insulation resistance is sufficient evidence that the windings are in condition to receive the High-pressure test. |
| *Machines for Tropical conditions* or other cases where the air temperature is in excess of 35°C (the permissible *Temperature rises* as above) *are to be reduced by* 20% except for Mica, Asbestos and fire proof materials. | | | | | | | Rated pressure / Resistance Megohm<br>Low pressure / 0,25<br>above 350 Volts / 1 |

## Overloads, Commutation Test, Starting.

a) Machines with *continuous rating* having limits of full-load temperature rise and *unobstructed ventilation* are to be capable of withstanding 25% *overload* for the periods:

| Outputs | Duration |
|---|---|
| 100 kW or HP and above | 2 hours |
| Below 100 kW or HP not below 25 kW or HP | 1 hour |
| Below 25 kW or HP not below 2 kW or HP | ½ hour |
| Below 2 kW or HP | 5 Min. |

b) Machines as above but *with partially obstructed ventilation* are to be capable of withstanding 15% *overload* for the periods as above.

c) Machines as above but of the *totally enclosed* class have *no overload rating* except momentary as required for commutation tests.

*Motors with short-time rating* are to be capable of carrying:

| Overload torque | Duration |
|---|---|
| 100% | 30 Secs |

### Commutation test.

D. C. machines must operate throughout the range from no load to the highest specified overload with fixed brush setting. The operation must be practically sparkless from no load to full load and without injurious sparking up to the maximum specified overload.

## Tolerances.

| Subject | Tolerance | |
|---|---|---|
| Speed of Shunt wound Motors | ±5% of speed at full load temp. | |
| Speed of Series wound Motors | ±7,5% of speed at full load temp. | |
| Speed variation of Asynchronous machines | 50% on the listed value of slip | |
| Inherent Pressure Regulation of Generators | 0,4 of specified value | |
| Efficiency | Full load Efficiency %*) | Tolerance %: |
| | ≧ 95 | 0,5 |
| | < 95 to 93 | 0,75 |
| | < 93 to 90 | 1 |
| | < 90 to 85 | 1,5 |
| | < 85 to 75 | 2 |
| | < 75 to 65 | 3 |
| | < 65 | 4 |
| Power Factor | Full load P. F. % | Tolerance %: |
| | ≧ 90 | 2 |
| | < 90 to 85 | 3 |
| | < 85 to 75 | 4 |
| | < 75 to 65 | 6 |
| | < 65 | 8 |
| Average load Guarantee for Efficiency | (1/1 load Efficiency × 4) plus (3/4 load ,, × 3) plus 1/2 load ,, × 2  divided by 9 | |
| | Tolerances as above | |
| Observation of Temp. Rise | 2% | |

*) Above values apply to conventional Efficiency. For directly measured efficiency the tolerance should be increased 1/2% for Efficiencies ≧ 88% and 1% for Efficiencies < 88%.

# SECTION II.

# Transformers.

### Rating-Classification.

| Germany | Britain | U. S. A. |
|---|---|---|
| a) *Continuous Rating.* The T. shall give its rated output for any desired period without exceeding the limits specified for observable temperature and temperature rise.<br><br>b) *Continuous Rating with short-time load.* The T. shall give the rated output for the specified period without exceeding the limits mentioned under a). The load period is assumed to be shorter than the time required by the T. to reach the constant end temperature condition; the no-load period (secondary windings disconnected) is assumed to be of such duration to enable the T. to reach the constant no-load temperature.<br><br>c) *Continuous Rating. Intermittent load.* Load periods of max. 5 min. are followed by no-load periods the duration of which is not sufficient to enable the T. to reach constant no-load temperature. The T. shall give the rated output with the specified relative load periods for any desired period without exceeding the limits mentioned under a). | a) *Continuous Rating* is the output at which a T. can work continuously for an unlimited period and comply with these rules. | a) *Continuous Rating.* The T. shall be able to operate continuously at its rated output, without exceeding any of the limitations established. |

## Continuation: **Rating-Classification.**

| Germany | Britain | U. S. A. |
|---|---|---|
| d) *Short-Time Rating.* The T. shall give the rated output for the specified period without exceeding the limits mentioned under a). | b) *Short-Time Rating* is the output at which a T. can work for a limited period (to be specified in each case) and comply with these rules. | b) *Short-Time Rating.* The T. shall be able to operate at its rated output during a limited period to be specified in each case without exceeding any of the limitations established. |
| e) *Duty-cycle Rating.* Periods under pressure of max. 5 Min. are followed by periods during which the T. is switched off (entirely), the duration of which is not sufficient to enable the T. to reach the temperature of the cooling medium. The T. shall give the rated load under the same conditions as mentioned under c). | c) *Duty-cycle Rating* (in special cases) is the continuous rating which is the thermal equivalent of the cycle of duty specified. | c) *Duty-Cycle Operation.* For purposes of rating either a continuous or a short-time equivalent load may be selected which shall simulate as nearly as possible the thermal conditions of the actual duty-cycle. |
| f) *Agricultural Rating.* 100 % overload is admissible for 12 hours per day and during approx. 500 hours per year. The T. shall give *160 % of the rated load* continuously without exceeding the limits mentioned under a) on *100 % overload* as above the specified limits for temperature and *temperature rise may be exceeded by 10° C.* | | d) *Nominal Rating.* (For special cases). The nominal rating of a substation transformer[*]) carrying traction load shall be the KVA-output at a stated power factor input, which, having produced a constant temperature in the transformer may be increased *50 % for 2 hours*, without producing temp. rises exceeding by more than 5° C the limiting values stated hereafter. These T.s should be capable of carrying *100 % overload for one minute* without disqualifying them for continuous service. |
| | | e) *Rating of Protective Reactors.* P. R. shall be rated by: <br> 1. KVA absorbed by normal current. <br> 2. Normal current, frequency as lim (delta) voltage. <br> 3. Short circuit current which the P. R. is required to stand. |
| | | [*] also for Substation machines. |

## Heating

| | Germany | Britain | U. S. A. |
|---|---|---|---|
| **a) Standard- and max. permissible Temperatures of Cooling Media.** | | | |
| Normal | 20° C | — | — |
| Max. permissible *air* | 35° C | 40° C | 40° C |
| Max. permissible *water* | 25° C | — | 25° C |
| Max. permissible *oil* | 95° C | (provisionally adopted) 90° C | 90° C |
| **b) Determination of End-Temperatures and Temperature Rises. Heating Test.** | | | |
| The *Heating Test* shall be carried out at rated load with the exception of T.s with agricultural rating: <br><br> 1. *For T.s with continuous rating* until the attainment of a steady Temperature. (Test started with T. warm or cold.) <br><br> 2. *For T.s with continuous rating and short-time load.* The test shall be started with the T. at constant no-load temperature and carried on for the specified period. <br><br> 3. *For T.s with short-time rating.* The test shall be started with cold Transformer i. e. with the temperature of the windings within 3° C (max. higher) of the temp. of the cooling medium and carried on for the specified period. <br><br> 4. *For T.s with intermittent ratings as under c and e on pages 31 & 32.* The T. shall be subjected to an intermittent load of the specified relative load periods and carried on until the attainment of a steady temperature. It shall be discontinued at the end of the last load period. For the test the duration of a load-cycle shall be 10 Minutes. <br><br> **Note:** *For Ratings a, c and e (pag. 31 & 32)* the Heating Test may be considered satisfactory if the *rate of increase of temperature* does not exceed $1°C$ *per hour* and *if the temp. is $5°C$ below the guaranteed value.* <br><br> 5. *For T.s with agricultural rating.* For the purposes of Heating Test the 60% overload is considered as rated load. The 100% overload heating test is started with the Transformer on constant rated load temperature and continued until the attainment of a steady temperature but not longer than 12 hours. | As one page 7. <br><br> *Air blast Transformers Correction of observed Temp. Rise:* <br><br> The Temp. Rise as stated on page 36 & 37 shall be reduced by 1 per cent for each 2° C that the temperature of the air entering the Transformer is below 40° C. | As on page 7. <br><br> Further: <br><br> *Air-blast Transformers. Correction of observed temp. rise* due to difference in resistance when temp. of ingoing air $t$ differs from that of the standard of reference i. e. 40° Cels. <br><br> $$\frac{274{,}5}{234{,}5 + t}.$$ <br><br> *Loading of T.s for Temperature Test.* <br><br> The conditions shall be such to give losses as nearly as possible equal to those obtained under normal or specified load conditions. |

## Continuation: **Heating.**

| Germany | Britain | U.S.A. |
|---|---|---|

### c) Methods of measurement of Temperature of the cooling media.

| Germany | Britain | U.S.A. |
|---|---|---|
| For 1. Transformer with air-cooling (self-cooling).<br>For 2. Transformer oil-immersed, self-cooling.<br>For 3. Transformer oil-immersed with separate oil cooler (air cooled) and forced oil circulation.<br>The temp. of the cooling medium is to be taken as the mean of the ambient air temperatures during the last quarter of the test period.<br>For 4. T.s of air blast type.<br>For 5. T.s oil-immersed, tank forced air cooled.<br>For 6. T.s oil-immersed, tank forced air cooled also forced oil circulation.<br>For 7. T.s oil-immersed with separate oil cooler which itself is forced air cooled and forced oil circulation.<br>The temp. of the cooling medium is to be taken as the mean of the temperatures of the air at the air inlet taken during the last quarter of the test period.<br>*For water cooled types:*<br>8. With individual parts only cooled. Windings not immersed.<br>9. Oil-immersed, cooler inside of tank.<br>10. Oil-immersed, cooler outside of tank and forced oil circulation.<br>The temp. of the cooling medium shall be taken as the mean of the temperatures of the water at the water inlet taken during the last quarter of the test period.<br>If the amount of heat carried off by the ambient air is considerable, the temp. of the cooling medium is to be considered the mean value derived from the formula<br>$$T_M = \frac{T_K W_K + T_L W_L}{W_K + W_L}$$<br>$T_M$ = mean value of temp. of cooling medium.<br>$T_L$ = temp. of ambient air.<br>$T_K$ = temp. of other cooling medium.<br>$W_L$ = heat carried off by ambient air kW.<br>$W_K$ = heat carried off by other medium kW. | The air temperature shall be taken as the mean of the readings of the thermometers taken at equal intervals during the last quarter of the duration of the Test. For pipeventilation or forced-draught, the air temp. is to be measured by a thermometer placed in the current of the incoming air. | The value to be adopted for the ambient temp. during a test, is the mean of the readings of the thermometers taken at equal intervals during the last quarter of the duration of test.<br><br>*Water-cooled T.s.* The temp. rise shall be based entirely upon the temp. of the cooling water. If under assumed standard conditions (water 25°, air 40°C) the amount of heat carried off by the air is 15% or more of the total, the temp. of the cooling water during test should be maintained within 5°C of that of the surrounding air. Where this is impracticable the ambient temperature should be determined from the change in the resistance of the windings, using a disconnected Transformer, supplied with the normal amount of cooling water until the temp. of the windings has become constant. |

## Transformers.

## Continuation: **Heating.**

| Germany | Britain | U.S.A. |
|---|---|---|
| <td colspan="3" align="center">**d) Methods of measurement of Temperature rises.**</td> |||
| The temperature rise of windings of *air cooled t.s* is defined as *the higher of* the *two values* as below<br>1. *Mean temperature rise by Resistance Method.*<br>2. *Temperature rise of the hottest accessible spot measured by Thermometer.*<br>For *oil immersed T.s* Method 1 only shall be used.<br>For *T.s for very heavy currents* the Resistance method is not exact; special arrangements for measuring oil temperature only are recommended. | *Resistance Method* for windings | The temperature of the windings shall be measured *by their increase of resistance and by thermometers* whichever measurement yields the higher temp., that temp. shall be taken as the highest observable temperature by method 2 (page 10). |
| *Temp. rise of core* shall be determined by Thermometer Method at the hottest accessible spot. | *Temp. rise of core* same as Germany. | *Temp. rise of core* same as Germany. |
| *Temp. rise of oil* shall be determined by Thermometer measured near oil surface in the tank. | *Temp. rise of oil* shall be determined by thermometer in oil. | *Temp. rise of oil* same as Germany. |
| <td colspan="3" align="center">**e) Correction of Temperature Rise for Altitude.**</td> |||
| As on page 11. | As on page 11. | As on page 11 but water-cooled T.s are exempt. |

3*

## f) Permissible Temperature Rises. (ET = End Temperature; TR = Temp. Rise.)

| Item | Germany | | | | Britain | | | | | U.S.A. | | | |
|---|---|---|---|---|---|---|---|---|---|---|---|---|---|
| | Class of Insulation | | ET °C | TR °C | Type | Insulation | | ET °C | TR °C | Type of Transformer | Class of Insulation | ET °C | TR*) °C |
| 1 | Fibrous material as paper, cotton, silk, wood | not impregnated | 85 | 50 | Air cooled | cotton, silk, paper and similar material | *not* impregnated or oil-immersed | 85 | 45 | Air cooled or air blast | Class O | 80 | 40 |
| 2 | | not impregnated but coil dipped in insulating fluid without pressure or vacuum | 85 | 50 | | | impregnated Class A | 95 | 55 | Air cooled or air blast | Class A same as Germany Items 2 & 3 | 95 | 55 |
| 3 | | impregnated or compounded i. e. impregnating material fills the space between conductor and insulation and between the fibres | 95 | 60 | | enamelled wire | Class A | 95 | 55 | | | | |
| | | | | | | Asbestos Mica etc. | Class B | 115 | 75 | | | | |
| 4 | | immersed in oil | 105 | 70 | Oil cooled | insulating materials immersed in oil | | 95*) | 55 | Oil cooled | Class A | 95 | 55 |
| 5 | Mica- and Asbestos preparates | | 115 | 80 | | | | | | — | — | — | — |
| 6 | Raw Mica, Porcelain or other fire-proof materials | | 5°C more than Items 1—5 | | | | | | | Water cooled | Class A | 80 | 55 |

*) based or water of 25°C

# Transformers.

| | Part of Transformer | | | | |
|---|---|---|---|---|---|
| 7 | Single-layer windings not insulated | 5° C more than Items 1—5 | | | |
| 8 | Continuously short circuited windings | as other windings by resist. method | | | |
| 9 | Core in | air cooled types (windings not immersed) | 95 | 60 | |
| 10 | Core in | oil-immersed types | 105 | 70 | |
| 11 | Oil near surface | | 95 | 60 | |
| 12 | All other parts | | restricted by influence on adjacent parts only | | |

*) Method of measurement: Items 1—6. Resistance Method. „ 7—12. Thermometer „

Note: The end temperature must under no circumstances be exceeded. The temp. rise may be exceeded only in such cases where the temp. of the cooling medium is *always* so low that the end temperatures (as above) will not be reached.

## Iron cores

| Part of Transformer | | | |
|---|---|---|---|
| | | restricted by influence on adjacent parts only | 50 |
| oil | | 90 | |

*) Method of measurement: see page 35.

| Part of Transformer | | ET | TR |
|---|---|---|---|
| Core | air cooled type | restricted by influence on adjacent parts only | |
| | oil cooled type | | |
| Oil | | 90 | — |
| All other parts | | restricted by influence on adjacent parts only | |

*) Method of measurement: see page 35.

## I. High-Pressure Tests.

| Nature of Tests | Germany | Britain | U.S.A. |
|---|---|---|---|
| | 1. High-Pressure Test<br>2. Surge Test for T,s from 1000 to 60000 Volts<br>3. Test on Turns | 1. High-Pressure Test<br>2. Insulation Resistance Test | 1. Test of Dielectric Strength<br>2. Insulation Resistance Test |

### 1. High-Pressure Test / Test of Dielectric Strength

| | Germany | Britain | U.S.A. |
|---|---|---|---|
| Wave form | sinusoidal | sinusoidal | sinusoidal |
| Frequency | rated, or 50 cycles | in general any between 25 and 100 rated | not less than rated 60 cycles recommended |
| Duration | one minute | one minute | one minute |
| Temperature during Test | working Temperature | working Temperature | working Temperature |

### Germany

| Type of Transformer | Rated terminal pressure E | Test pressure Volts | min. Volts |
|---|---|---|---|
| All | up to 10000 Volts | 3,25 E | 2500 |
| | over 10000 Volts | 1,75 E + 15000 | — |
| Air cooled (windings not immersed) | up to 10000 Volts | if measured cold: 3,74 E | 2800 |
| | over 10000 Volts | if measured cold: 2,02 E + 17250 | — |

### Britain

| Type | Rated Output | Test pressure Volts |
|---|---|---|
| All | below 746 Watts | 2 E + 500 |
| | 746 Watts and above | 2 E + 1000 |
| Transf. in service (not new) | below 746 Watts | 1,5 E + 375 |
| | 746 Watts and over | 1,5 E + 750 |

### U.S.A.

| Type of Transformer | Rated output Normal circuit voltage E | Test pressure Volts |
|---|---|---|
| | for 25 Volts or lower | 500 |
| Household devices | not over 660 Watts<br>„ „ 275 Volts | 500 |
| Standard | all | 2 E + 1000 |
| Distributing Transformers | from 550 to 4500 Volts | primary: 10000<br>secondary: 2 E + 1000 |

| Transformer Bushings | up to 3000 Volts | 8 E + 2000 | — | Transf. for direct connection to consumers circuit | primary less than 2000 Volts | — | | A. C. single phase permanently grounded (not 3-phase with earthed neutral) | over 300 Volts | 2,73 E' E' + 1000 voltage of circuit to ground |
|---|---|---|---|---|---|---|---|---|---|---|
| | over 3000 Volts | 2 E + 20000 | — | | primary 2000 Volts and above | 5000 | 10000 | | | |
| | | | | | secondary (L.P.) windings | | $2E + 1000$ | | | |
| | | | | Auto Transformer | | — | same as apparatus to which connected | Auto Transformer | — | same as apparatus to which connected |

E represents:

1. The rated terminal pressure when testing individual windings against core.
2. For Current and regulating T.s with separate primary and secondary windings, the rated terminal pressure of the circuit to which the windings are connected in series.
3. For series connected windings, the sum of the terminal pressures.
4. For regulating T.s with variation of secondary voltage by switching off and on of additional windings, the primary terminal voltage necessary for max, secondary voltage.
5. For T.s with one outer pole permanently grounded 1,1 times rated terminal pressure.

**Note**: At constant voltage the current must not increase continuously nor must there be sudden fluctuations.

**Note**:
1. The test must be based on the highest pressure E to which the windings may be subjected.
2. T.s are to have the same test between high pressure winding and core as between high pressure winding and low pressure winding.

**Note**:
E represents the normal voltage of the circuit to which the machine is connected.

## Transformers Section II.

### 2. Surge Test (Germany only).

The surge test (Purpose see under Section I) shall be made on the factory test bed with the completely assembled transformer with windings for rated voltages from 25 to 60 kV in accordance with the diagrams of connections as shown below.

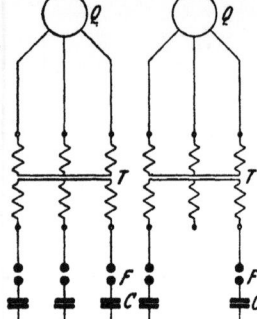

The Transformer winding on test shall be connected to cables or condensers (C) over sphere sparkgaps (F) with solid spheres of at least 50 mm diameter. — The test capacitance of the cables or condensers shall be as follows:

| Rated Voltage E kV | Capacitance C of each phase min. $\mu$ F. (Mikrofarads) | Suitable form of capacitance |
|---|---|---|
| 2,5 up to 6 | 0,05 | cable or condenser |
| ,, ,, 15 | 0,02 | ,, ,, ,, |
| ,, ,, 35 | 0,01 | ,, ,, ,, |
| ,, ,, 60 | 0,005 | condenser |

Details of test as for machines Section I.

| 3. Test on Turns. | 2. Insulation Resistance. | 2. Insulation Resistance Test. |
|---|---|---|
| Germany | Britain | U. S. A. |
| The winding test shall be made under no-load condition by increasing of the supplied voltage up to the values given below. The frequency can be increased correspondingly.<br><br>Duration of Test: 5 Minutes. | same as U.S.A. | The insulation test may afford a useful indication as to whether a machine is in suitable condition for application of the dielectric test. The insulation test shall be made with all circuits of equal voltage above ground connected together. Circuits or groups of circuits of different voltage above ground shall be tested separately.<br><br>*Test pressure if possible D. C. 500 Volts.*<br><br>Minimum values in Megohms for dry apparatus (not oil-immersed) see page 17. |

| Output kVA | Test voltage / Rated voltage |
|---|---|
| up to 1000 | 2 |
| over 1000 | If possible 2; min. 1,3 |

### Working in parallel (Germany only).

It is recommended that T.s of different outputs and with a ratio of outputs over 1 : 3 shall not be required to work in parallel continuously.

T.s shall be considered to work in parallel satisfactorily if the rated impedance voltages do not differ from their mean value by more than $\pm 10^0/_0$. T.s with tappings shall not in every case be required to work in parallel satisfactorily on all ratios.

### Overloads.

| In general *none specified* with the *exception of agricultural T.s* see page 32. | none specified | In general *none specified* with the *exception of nominal rating* see page 32. |
|---|---|---|

## Efficiency, Transformer Losses.

| Germany | U. S. A. |
|---|---|

### 1. General.

| | |
|---|---|
| *Recognized Efficiencies:* Losses only mentioned, therefore Conventional Efficiency implicitly recognized only.<br>*Miscellaneous losses:* The power consumption of motors for separate ventilating blowers as also that of water- or oil-circulating pumps shall be stated separately.<br>*General conditions:* see page 20. | *Recognized Efficiencies:*<br>1. Directly measured Efficiency $= \dfrac{\text{Output}}{\text{Input}}$<br>2. Conventional Efficiency (determined from losses).<br><br>*General conditions:* see page 20. |

### 2. Classification of Losses and their Measurement.

| | |
|---|---|
| *No-load Losses* are considered equivalent to the input at normal rated pressure on the primary side, at rated frequency and with the secondary terminals disconnected. They include the core loss, the dielectric loss and the $I^2R$ loss due to the no-load current. The n. l. l. are generally measured from secondary side. | *No-load Losses* include the core loss, the $I^2R$ loss due to the exciting current and the dielectric loss in the insulation.<br>They shall be measured with open secondary circuit at the rated frequency, and with an applied primary voltage giving the rated secondary voltage plus the IR drop which occurs in the secondary under rated load conditions. |
| *Load Losses* are considered equivalent to the $I^2R$ losses at rated current and frequency, at working temperature measured within the terminals.<br>They shall be measured by applying the impedance voltage to the T. with the secondary windings short circuited. Stray-load losses due to eddy-currents are included in the losses measured as before. | *Load Losses* include $I^2R$ losses and stray load-losses due to eddy-currents caused by fluxes varying with the load.<br><br>They shall be measured by applying a primary voltage, at rated frequency, sufficient to produce rated load current in the with the windings secondary windings short-circuited. |

### Inherent Regulation

to be measured at the temperature attained under normal operation.

| | | |
|---|---|---|
| The I. R. of a T. at a specified Power Factor is the rise of the secondary voltage in per cent of the rated secondary pressure if the load is altered from rated load to no load and with primary voltage and frequency constant. | **Britain**<br>—  | **U.S.A.**<br>Same as Germany. Unless a P. F. is specified the regulation is understood to refer to P. F. = 1. |

### Tolerances.

| Germany . . . | for losses | $10\%$ of copper and core losses. |
|---|---|---|
| Britain . . . . | for temp. rise | expressly none. |

## Appendix to Section II (Transformers).
British Electrical and Allied Manufacturers Association Rules.
Fourth Edition 1920.

### Rating Classification.
a) continuous.
b) short-time.

### Heating.
a) *Standard- and max. permissible temp. of cooling media.*
b) *Determination of End-Temp. and Temp. Rises.*
c) *Methods of measurement of temperatures of cooling media.*

  } same as Appendix to Section I.

d) *Methods of measurement of Temperature Rises:*
  oil cooled types: Resistance Method,
  air   „    „  : Thermometer „
  Temp. Rise of oil measured by thermometer as also core temperature.
e) *Correction for Altitude* same as Appendix to Section I.
f) *Permissible Temp. Rises.*

| Class of Insulation | Type of Transformer | E T °C | T R*) °C |
|---|---|---|---|
| Fibrous material as paper, cotton, silk | air-cooled (generally) | 85 | 50 |
| | oil cooled | 85 | 50 |
| Mica and Asbestos preparates | | Higher values permissible (not specified) | |
| Raw Mica or other fire-proof material | | | |

\*) Method of measurement see under d).

**Note:** *Transformers for tropical conditions* or other cases where the air temp. is in excess of 35°C the permissible *temp. rise* (as above) is *to be reduced* by 20% except for Mica or Asbestos Insulation.

### High Pressure Test and Insulation Resistance.
Nature of Tests etc.: same as B.E.S.A.

1. *High Pressure Test.*

| Rated terminal pressure E | Test pressure Volts | min. Volts |
|---|---|---|
| not more than 333 Volts | 1000 | — |
| above 333 Volts not more than 500 Volts | 3 E | 1500 |
| Above 1500 Volts but not more than 2250 Volts | 4500 | — |
| above 2250 Volts | 2 E | — |

Note: same as B.E.S.A. page 38 & 39.

2. *Insulation Resistance.* In general the following insulation resistance is sufficient evidence that the windings are in condition to receive the high-pressure test.

| Rated pressure | Resistance megohms |
|---|---|
| low pressure | 0,25 |
| above 350 Volts | 1 |

### Overloads.
Overload capacity:
  25% for 2 hours,
  100% for 30 secs.

### Tolerances.

| Full-load Efficiency (by summation of losses) | Tolerance |
|---|---|
| 98% or higher | 0,25% |
| below 98% | 0,4% |
| average load | see page 30 |

Verlag von Julius Springer in Berlin W 9

**Erläuterungen zu den Vorschriften für die Konstruktion und Prüfung von Installationsmaterial, den Vorschriften für die Konstruktion und Prüfung von Schaltapparaten für Spannungen bis einschl. 750 V und den Normalien über die Abstufung von Stromstärken und über Anschlußbolzen.** Im Auftrage des Verbandes Deutscher Elektrotechniker herausgegeben von **Georg Dettmar**, Generalsekretär des Verbandes. Mit 46 Textabbildungen. Unveränderter Neudruck. 1922.
GZ. 3,6; 3,60 Fr.; 0,75 $; 3 sh.

**Erläuterungen zu den Normalien für Bewertung und Prüfung von elektrischen Maschinen und Transformatoren, den normalen Bedingungen für den Anschluß von Motoren an öffentliche Elektrizitätswerke und den Normalien für die Bezeichnung von Klemmen bei Maschinen, Anlassern, Regulatoren und Transformatoren.** Im Auftrage des Verbandes deutscher Elektrotechniker herausgegeben von **Georg Dettmar**. Sechste Auflage. Erscheint 1923.

**Erläuterungen zu den Vorschriften für die Errichtung und den Betrieb elektrischer Starkstromanlagen** einschließlich Bergwerksvorschriften und zu den Merkblättern für Starkstromanlagen in der Landwirtschaft. Im Auftrage des Verbandes Deutscher Elektrotechniker herausgegeben von Geh. Reg.-Rat Dr. **C. L. Weber**. Dreizehnte, vermehrte und verbesserte Auflage. 1923. GZ. 3.3; 4 Fr.; 0,80 $; 3 sh. 4 p.

**Arnold-la Cour. Die Wechselstromtechnik.** Herausgegeben von Professor Dr.-Ing. **E. Arnold**, Karlsruhe. In fünf Bänden. Unveränderter Neudruck. Erscheint im Frühjahr 1923.

I. **Theorie der Wechselströme.** Von J. L. la Cour und O. S. Bragstad. Zweite, vollständig umgearbeitete Auflage. Mit 591 Textfiguren.

II. **Die Transformatoren.** Ihre Theorie, Konstruktion, Berechnung und Arbeitsweise. Von E. Arnold und J. L. la Cour. Zweite, vollständig umgearbeitete Auflage. Mit 443 Textfiguren und 6 Tafeln.

III. **Die Wicklungen der Wechselstrommaschinen.** Von E. Arnold. Zweite, vollständig umgearbeitete Auflage. Mit 463 Textfiguren und 5 Tafeln.

IV. **Die synchronen Wechselstrommaschinen.** Generatoren. Motoren und Umformer. Ihre Theorie, Konstruktion, Berechnung und Arbeitsweise. Von E. Arnold und J. L. la Cour. Zweite, vollständig umgearbeitete Auflage. Mit 530 Textfiguren und 18 Tafeln.

V. **Die asynchronen Wechselstrommaschinen.**
1. Teil: **Die Induktionsmaschinen.** Ihre Theorie, Berechnung, Konstruktion und Arbeitsweise. Von E. Arnold und J. L. la Cour unter Mitarbeit von A. Fraenckel. Mit 307 Textfiguren und 10 Tafeln.
2. Teil: **Die Wechselstromkommutatormaschinen.** Ihre Theorie, Berechnung, Konstruktion und Arbeitsweise. Von E. Arnold, J. L. la Cour und A. Fraenckel. Mit 400 Textfiguren, 8 Tafeln und dem Bildnis E. Arnolds.

*Die Grundzahlen (GZ.) entsprechen den ungefähren Vorkriegspreisen und ergeben mit dem jeweiligen Entwertungsfaktor (Umrechnungsschlüssel) vervielfacht den Verkaufspreis. Über den zur Zeit geltenden Umrechnungsschlüssel geben alle Buchhandlungen sowie der Verlag bereitwilligst Auskunft*

Verlag von Julius Springer in Berlin W 9

**Theorie der Wechselströme.** Von Dr.-Ing. **Alfred Fraenckel.**
Zweite, erweiterte und verbesserte Auflage. Mit 237 Textfiguren. 1921.
Gebunden GZ. 11; gebunden 11,35 Fr.; gebunden 2,30 $; gebunden 9 sh. 6 p.

**Ankerwicklungen für Gleich- und Wechselstrommaschinen.**
Ein Lehrbuch. Von Prof. **Rudolf Richter,** Karlsruhe. Mit 377 Textabbildungen. Berichtigter Neudruck. 1922.
Gebunden GZ. 11; gebunden 14 Fr.; gebunden 2,80 $; gebunden 11 sh. 9 p.

**Kurzes Lehrbuch der Elektrotechnik.** Von Dr. **Adolf Thomälen,**
a. o. Professor an der Technischen Hochschule Karlsruhe. Neunte, verbesserte Auflage. Mit 555 Textbildern. 1922.
Gebunden GZ. 9; gebunden 9,60 Fr.; gebunden 1,95 $; gebunden 8 sh.

**Die wissenschaftlichen Grundlagen der Elektrotechnik.**
Von Prof. Dr. **Gustav Benischke.** Sechste, vermehrte Auflage. Mit 633 Abbildungen im Text. 1922.
Gebunden GZ. 15; gebunden 18 Fr.; gebunden 3,60 $; gebunden 15 sh.

**Hilfsbuch für die Elektrotechnik.** Unter Mitwirkung namhafter Fachgenossen bearbeitet und herausgegeben von Prof. Dr. **Karl Strecker,** Geh. Oberpostrat. Zehnte, umgearbeitete Auflage. In drei Teilen.
In Vorbereitung.

**Elektrische Starkstromanlagen.** Maschinen, Apparate, Schaltungen, Betrieb. Kurzgefaßtes Hilfsbuch für Ingenieure und Techniker sowie zum Gebrauch an technischen Lehranstalten. Von Studienrat Dipl.-Ing. **Emil Kosack,** Magdeburg. Sechste, durchgesehene und ergänzte Auflage. Mit 296 Textfiguren. 1923. GZ. 5; gebunden GZ. 5,8; 6,25 Fr.; gebunden 7,25 Fr.; 1,25 $; gebunden 1,45 $; 5 sh. 3 p; gebunden 6 sh.

**Schaltungen von Gleich- und Wechselstromanlagen.** Dynamomaschinen, Motoren und Transformatoren, Lichtanlagen, Kraftwerke und Umformerstationen. Ein Lehr- und Hilfsbuch. Von Dipl.-Ing. **Emil Kosack,** Studienrat an den Staatl. Vereinigten Maschinenbauschulen zu Magdeburg. Mit 226 Textabbildungen. 1922. GZ. 4; gebunden GZ. 6; 6,80 Fr.; gebunden 10 Fr.; 1.75 $; gebunden 2 $; 5 sh. 8 p; gebunden 8 sh. 3 p.

**Der Drehstrommotor.** Ein Handbuch für Studium und Praxis. Von Prof. **Julius Heubach,** Direktor der Elektromotorenwerke Heidenau G. m. b. H. Zweite, verbesserte Auflage. Mit 222 Abbildungen. 1923.
Gebunden GZ. 14,5; gebunden 18 Fr.; gebunden 3.60 $; gebunden 15 sh.

**Elektrotechnische Meßinstrumente.** Ein Leitfaden. Von **Konrad Gruhn,** Oberingenieur und Gewerbestudienrat. Zweite, vermehrte und verbesserte Auflage. Mit 321 Textabbildungen. 1923.
Gebunden GZ. 5,8; gebunden 6,40 Fr.; gebunden 1.30 $; gebunden 5 sh. 6 p.

*Die Grundzahlen (GZ.) entsprechen den ungefähren Vorkriegspreisen und ergeben mit dem jeweiligen Entwertungsfaktor (Umrechnungsschlüssel) vervielfacht den Verkaufspreis. Über den zur Zeit geltenden Umrechnungsschlüssel geben alle Buchhandlungen sowie der Verlag bereitwilligst Auskunft.*

MIX
Papier aus verantwortungsvollen Quellen
Paper from responsible sources
FSC® C105338

If you have any concerns about our products,
you can contact us on
**ProductSafety@springernature.com**

In case Publisher is established outside the EU,
the EU authorized representative is:
**Springer Nature Customer Service Center GmbH
Europaplatz 3, 69115 Heidelberg, Germany**

Printed by Libri Plureos GmbH
in Hamburg, Germany